EXCELLENT ONLINE TEACHING

Effective Strategies for a Successful Semester Online

AARON JOHNSON

ISBN: 0-9897116-0-9
ISBN-13: 978-0-9897116-0-9

www.excellentonlineteaching.com

Table of Contents

Preface

The One Thing You Need

Imagine that a close friend or perhaps your grown child moved to a foreign country. Distance would change the dynamics of your relationship, and communication would be difficult at times. But you would persist; you would figure out new ways to keep your friendship alive, and in the end, you would develop a whole new set of habits. I think teaching in the online environment is the same: it's about growing a relationship (a learning relationship), and it's about developing new habits to make that happen.

It's not about the technologies you choose--those are just tools. That doesn't mean learning the technical ropes is unimportant, but what the online teacher needs more than anything is to develop a set of habits: regular and clear communication, demonstrating compassion, and developing the discipline of prompt feedback. It's one thing to do these sometimes, but it's another for them to become routine. That's the core challenge of teaching online, and it's what separates the excellent online teacher from the just-okay one. It's the one thing you need: habit.

The residential classroom comes with built-in structures that reinforce our habits. We are provided schedules, office hours, even bells that ring to remind us of what's up next. The online environment can be disorienting because we lose this framework. So, our first task is to create new structures, structures that become the foundation for successful teaching habits. The great thing is we get to create them for ourselves. Many of the strategies and tips in this book are designed to help you build these new structures and habits.

Structures are the foundation for our new teaching habits, but there's a second essential ingredient. We need a sense of connection with our students. This is what fuels our habits in the midst of the mundane. Many teachers avoid teaching online courses because they fear they will lose touch with their students, that there will be absolutely no sense of relationship. The truth is that you just have to be more intentional about cultivating those relationships. I think that's a good thing.

Let's go back to the scenario at the beginning of this chapter. Our good friend moves to a foreign country, and at some point in that friendship we realize, if we want to remain connected, it's going to take some extra work. A lot of that will depend on how intentional we are with our communication. In your online course, that also means creating avenues of communication with your students where they have opportunities to ask questions and to share their lives with you. That connection is your fuel to excellence as an online instructor. One of the aims of this book is to give you starting points and strategies for developing this kind of connection with your learners.

To sum that up: It's all about teaching habits. Those habits are built on the structures you put in place, and your habits are fueled by the connection you develop with your learners.

If you are new to teaching online, I'll be honest and say that it may feel like you're starting over as a teacher. In a lot of ways, you are. Having taught for over a decade and worked with professors for the past five years, I know that you don't have time to pour over books full of theory (maybe during the summer you will). I love theory and abstract ideas, especially on the subject of education, but this book will be light on theory, and heavy on practices that get results. My hope is that you'll end the semester with a new level of confidence and with course evaluations that let you know that you are on the right track. More importantly, I hope that you and your students will be experiencing a vibrant learning relationship. If you have been teaching online for a while, I think you'll find some ways to improve your game.

The world of online education may be one of the most rapidly changing fields of our day. Because of that, this book will focus on more enduring topics, like habits, structures, and connection. Yet, there are important tools and innovations in our line of work, an unavoidable technical aspect to all of this. As you read, you'll notice that I refer to the book's companion website www.excellentonlineteaching.com. These links will supplement each chapter with resources and provide you with things like screencasts and updates on the more technical oriented topics. This gives a dynamic edge to the book without having to update it every two months.

The sixteen chapters in *Excellent Online Teaching* correspond with the sixteen weeks in a typical semester. This book is succinct, but it's dense. So, I recommend reading the entire thing, then go back and focus on one chapter at a time, giving yourself a week to hone those skills.

Finally, like any good online teacher, I'd like to make myself available. If you have any questions, feel free to contact me at aaron@excellentonlineteaching.com

Chapter 1

Am I Ready?
3 Ways to Prepare for Your Course

In the old westerns the hero has a common plan: to capture the bank robbers, he must take a little known route, a hidden shortcut enabling him and his posse to "head them off at the pass." That's what this chapter is about. As you prepare for your online class, it's crucial to head a few things off at the pass. I recommend that you block out an hour or two to survey your course and review these three critical areas.

1. Survey Your Course and Syllabus

Log in to your course and look it over. Here are six things to note:

- *What's going on the first week?* Click into every item in the first week of the course. This will put you into the shoes of your students and will help you as you build that first weekly email (see chapter 3). This is one of those important habits worth developing. I recommend doing this on a weekly basis as you prepare for the upcoming week of your course.

- *Where and when will you engage your students?* Because effective learning is interactive learning, note discussions and other places in the course where feedback will be most important during your semester or quarter.

- *Is there some element or set of instructions on the course site that you find confusing?* If it's confusing to you, then expect to get emails from your students. Give special attention to more complex assignments. Would students benefit from an example paper, a grading rubric, or an FAQ page? If so, put those on your to-do list.

- *Review your syllabus.* After teaching his first online course, a professor whom I work with said that the best advice he could give to online teachers was to "read, reread, and then reread your syllabus [again]." Try to see your course through the eyes of your students. This may be difficult, so bring in another set of eyes to help you out: a friend, your spouse, or teaching assistant.

- *What are the most demanding weeks in your course?* Note the weeks when major assignments are due. Are the exams automatically graded by your learning management system, or will they require a greater time investment to grade? Are there two assignments due the same week that could be placed in separate weeks? The pacing of your course should be noticeable and reliable for your students.

- *Are there assignments that build upon a previous assignment?* If so, timely feedback will be critical to the learning process for your students. Mark those on your calendar so that your grading and feedback don't get boxed out by other demands.

2. Your Communication

- *What is your preferred mode of communication?* Is it email, chat, social media, phone, or perhaps a course discussion board? In your early correspondence with your students, let them know how you plan to communicate with them. I worked with an online instructor who was happy to engage students in lively email conversations, but most of his students weren't aware of this and never took advantage of it. Two things are important here: 1) students will follow the model of communication you create, and 2) invitation is the key to accomplishing this.

For example, I'd encourage you to write to them during the first week of the course and say:

> *"I invite you to email me questions that come up as you work through the reading, to push back, and argue your viewpoint."*

- *Invite Requests for Feedback.* I can't tell you how many times I've heard this from students' course evaluations (including my own), "I didn't get the feedback that I hoped for on my assignments." That's an easy one to head off at the pass through the same strategy of invitation. You might write:

"If you ever find that you didn't get the feedback that you needed on an assignment, then email me. Just be sure to ask me specific questions about specific parts of your assignment. This is an important part of the learning process, so engage me on this."

This reminds your students that they play an active part in the learning process. One-time invitations won't be enough; you'll need to send your learners a similar message again when you return graded assignments. I think this brings up an important reality of online education: you're going to have to over-communicate. (This is why *Evite* sends out all those reminders and opportunities to respond!) It doesn't mean nagging our students, but they might interpret it that way. Chapter three will give you some communication tools for navigating effective communication.

- *Email and Turn-Around Time Expectations.* Let your students know what kind of turn around they should expect from you. I strongly recommend you reply to emails within 24-48 hours. If you don't plan to reply to email on the weekends, or on particular days, then let your students know.

How long should they expect for you to grade and return assignments? I recommend 1 to 2 weeks, but a lot of this depends on the nature of the assignment and the size of your course. Your institution may have standards for responding to student emails and for returning assignments. Turning these into habits will be crucial, if not the most important part, to getting positive student evaluations. Be realistic, though; under-promise and over-deliver.

3. Your Calendar

There will be weeks in the semester that are more demanding than others. Note exam weeks and when papers are due. Reserve space in your calendar for those learning activities and assessments that will require more time.

Will you be out of town, out of the country, or have limited access to the internet during the semester? Let your students know at the beginning of the course, and let your institution's support team know.

Completing this review of your course site, syllabus, communication, and calendar will save you hours this semester and set you up for a great experience.

Chapter 2

Where Do I Begin?
How to Get Your Online Course Off to a Great Start

Part of becoming an excellent online teacher is being an effective one. As I'm sure you know, there are teaching practices that yield more payoff than others. Here I want to let you in on two practices that will yield you the most payoff during the first week of your course.

1. Send Out a Welcome Message.

This sets the stage and tone for the entire course. Your welcome message also makes students aware of your presence; it communicates that you're not a course supervisor, but someone who intends to be actively involved in facilitating their learning process. Your course management system should have a feature that sends a mass email out to all of your students. If you haven't located this feature, contact your technical support team and get the low-down on how to use it. This is one of those indispensable tools in your online teaching toolbox.

Five Elements for Crafting an Excellent Welcome Message

- An Attention-Grabbing Headline

Make your subject line something that grabs their attention and gets them to read. I'd recommend stealing the headline/title from this chapter. "How to..." headlines are a reliable formula for email subject lines.

See the chapter two resources at www.excellentonlineteaching.com for more headline tips.

- Encouragement

Take a moment to consider what's going on in the lives of your students this week. Many have just finished a long semester and are starting a new one. For some, this may be their first online course, and they're daunted at the prospect of trying something new. Recognizing their hard work and how they are feeling will go a long way in communicating that you care about your students and want to connect with them.

- 2 Things

After reviewing the course, what are the two most important things your learners need to know? Do they have a discussion post or some other initial assignment due soon? Is there a part of the reading you want to draw their attention to or a question for them to consider as they read? Online students can freeze during the first week of the course just because they can't figure out where to begin. This happened to me recently in a course that I took on hybrid course design, and it gave me some insight into what our student experience. It was a great course, but I just couldn't figure out where to begin. The 2 Things approach can be a great way to set your learners at ease by giving them a clear starting point.

- DTR (Define the Relationship)

Tell them about your role in the course. Invite them to contact you (put your contact info in the email even though it is accessible elsewhere). Explain how you plan to participate in their threaded discussions. Let them know how you plan to engage their learning process.

- Expectations

This is a great place to relay those communication details mentioned in chapter one: reply times, feedback expectations, when you may be traveling during the semester.

Now, this is a lot to include in a single welcome message. The fact is if your message is too long, then your students will stop reading. Because of this, you may want to break your welcome message into two. Also, be sure to take time to format your message into distinct parts by using heading and by keeping your paragraphs succinct. See the chapter two resources online for a welcome message example.

2. Engage Student Introductions

A course *introductions discussion* is a staple part of any online course. A great way to communicate that you are interested, attentive, and involved is to engage your students' introductions. Here are two ways to communicate all of that.

- Click into the *introductions discussion*, and type up a post with your responses to the same introduction questions your students are answering. Using copy & paste, post that same introduction to each group. Some courses may not use group mode for their discussions. If this is the case, you'll only need to post once to the discussion.

I recommend working with distinct groups of four to five students because it makes the course more manageable. This is one of those important "structures" that I mentioned in the introduction. If you are unfamiliar with how to create groups and send responses to distinct groups of students, contact your technical support team.

- Connect with your students. Go into the discussions midweek, when your students have had a chance to post their introductions. Read their posts, make brief comments, or ask brief questions. You may find that you grew up in the same part of the country, or share the same hobby. Keeping your responses short and sweet makes this doable.

One professor I work with writes individual emails to his students during the first week of the course. If you have a large class, this can be time-consuming, but it may be worth it. If you take this approach, I recommend creating an email template, then personalizing it as you work on each email.

The Payoff

These two strategies, sending out a welcome message and participating in the student introductions, inject the course with your presence, communicating that you are involved in your students' learning process and that you are available to them. When you make this happen, especially at the beginning of the course, you're creating the beginning of a vibrant learning experience.

Chapter 3

How Do I Communicate Well with My Students? 4 Strategies for Effective Online Communication

You're well aware that communicating online is quite different from a face-to-face conversation. For all you know, I just wrote that last sentence with a sarcastic tone; maybe I meant it to be lighthearted. Perhaps I was gravely serious? It's hard to tell because you can't read my non-verbals, and you can't hear the texture of my voice (which might have told you that I'm recovering from a cold).

You can see the communication gap and how it can easily create both misunderstanding and confusion. So how do you make up for that gap? Here are four practices that will go a long way to make your online communication clear and effective.

1. Use Emoticons and Use Them Often

You may be asking, "What are emoticons?" They're these guys: :o) ;) :0(

Okay, I'll admit, I used to judge emoticon users. In my mind, they were those annoying people who forwarded emails and watched kitten videos all day. But after having several students read a negative tone into my messages, I decided to try out the smileys - and they worked!

For some reason, our tendency as humans is to read things in a negative light. Emoticons help to mitigate that dynamic, and they help you to convey the more positive aspects of communication, such as encouragement, openness, and compassion. I'll put a smiley after anything that I think could be misinterpreted, and I never use negative emoticons when referring to student work. Your learning management system and email will likely have these built into their HTML editor. Leverage them.

2. Communicate Often

Imagine a face-to-face course where the professor shows up late for class. The first week she's five minutes late, the second week it's fifteen minutes. What would be the fallout? Her students would begin to disengage, and she would see empty seats marking the curious absence of several students. The same principle is at work in your online course: if your students feel like you are absent, they are more likely to disengage. When you model attention and presence, you should witness an increase in participation.

The easiest way to inject your presence into the course is to send out a weekly email on Monday morning. See chapter four for a weekly email recipe. Save copies of these messages, because you can reuse them next semester. Of course, you'll need to review and revise them, but your work can be repurposed.

3. Address Students by Name

The word we enjoy seeing and hearing most is our own name. When an instructor begins her correspondence or grading comments with our name, we take notice; it communicates that we are seen as a real person who matters. This is an easy way to make your students more receptive and more engaged, but it's an easy practice to forget. Make it one of your habits.

4. Read Carefully

I don't know how many times I've responded to a student without carefully reading the details of their email. This usually happens when I'm busy scanning my email, but really don't have time to respond. (Did you know that checking email can be as addictive as playing a slot machine? I've included an article about this on the chapter three resources page online.) One way to prevent this is to simply put it in your calendar. Set aside times just for checking and responding to student correspondence, and be sure to give their emails the same attention you want them to give to yours. Calendaring your email time is another one of those foundational structures that will aid you in creating effective teaching habits.

Chapter 4

How Do I Start Each Week? 5 Core Ingredients to Your Weekly Email

Assume that your students start each week lost, treading water in a sea of links and documents. It's your job to throw them a lifeline and to help them get oriented. The weekly email does just that. It's also one of the essential habits of an excellent online teacher. So, put it in your calendar for Monday morning.

5 Core Ingredients for a Great Weekly Email

1. "Where Do I Start?"

The first thing your students need at the beginning of the week is your answer to, "Where do I start?" This may seem obvious to you, but that's because you are familiar with the course. Review the course site for the week and look into the weeks ahead to see if there are any major assignments coming due. Try to anticipate questions your students may have for the current week's work and for upcoming major assignments. Use the *2 Things* method explained in chapter two.

2. Make your Subject Line Interesting

We all get flooded with emails, which means that you need to give your students a reason to read yours. Of course, that includes putting valuable content in your message, but you have to get them hooked first. Author Jeff Goins has written a fantastic article where he recommends the following five tips for writing effective headlines (in our case, subject lines).

- Use numbers: *Week 12: The 2 things you need to know.*
- Use interesting adjectives: *Easy Ways to Get Ahead this Week*
- Use unique Rationale: *3 Reasons to Start on Your Paper Now*
- Use "What, Why, How, or When: *Why you want to begin your interviews now*

- Make and Audacious Promise: *How to Get an A on Your Interview Assignment*

I highly recommend Goin's full article, *5 Easy Tricks to Help You Write Catchy Headlines.* You can find a link to this article in the chapter four resources online.

3. Provide Rationale

Why is this reading important? How does this assignment prepare me for my future or current career? As more and more Millennials (also known as Generation Y) enter graduate education, instructors are realizing just how important it is to provide the rationale for assignments. These students want to know how the content connects with everyday life.

Why are we reading this? Why are we doing this? These kinds of statements strike some instructors as just plain annoying, but there is a good reason for using them. Millennials have grown up in an information-rich culture where they are constantly required to make decisions about what information is worth their time. When you explain the significance of coursework, you're connecting with students who have more of a pragmatic learning style.

Moreover, when you provide rational, you're modeling integration. You are showing students how the course content impacts daily life and prepares us for our vocation. You are helping build interest in the coursework, which changes a student's attitude going into it. However, don't feel like you have to write a paragraph-long justification for your assignments--that's not what it's about. It could be a succinct sentence or two, or a question: *"You'll find the journal article assigned for this week has some interesting professional implications. As you read, note how this might help you in a counseling* (or other professional) *situation."* The significance of the reading, and your rationale for assigning it is implied within those brief instructions.

4. Communicate Your Expectations

Ask yourself, "What am I *most* looking for from my students this week or in this particular assignment?" Then let them know. If you've taught the course before, you are likely aware of the errors and misunderstandings that are common when students are working through the assignments and material for the week. Note these as you review your course each week and point them out.

5. Be Positive & Encouraging

Begin and end your message with encouragement; maybe it's a lighthearted anecdote from your weekend or something you are enjoying about your students.

Being positive isn't always easy. I'll be quick to admit that it's easy to create a negative portrait of our online students. This was especially true during my first years of teaching online. I could feel my attitude shift while grading some low-quality papers and receiving a few of those you-could-have-checked-the-syllabus-instead-of-asking-me emails. That negative frame of mind crept into the tone of my course communications. Because online students need a higher dose of encouragement, this negative portrait was incredibly counterproductive.

From numerous conversations with colleagues, I found that this wasn't just my own negative bent; it seems to be a very normal experience for online educators. Here are a few things I've found that will preempt negativity and help create a more positive and accurate view of my students.

- Take a break. When my tone gets negative, it's usually when I'm tired or have been at the computer for a long time. Get up and go for a walk, take a nap, or grab a bite to eat. You'll come back to your work with a new outlook.

- Review your course participants list. As I see their photos and reflect on their lives, I begin to see my students as real people. This jogs my memory that Jill just lost a close family member, that Charlie has been sick, and that Leslie did an amazing job interacting with her group last week. This is a really effective way to get perspective.

- End your email on a positive note, framing the week for your students by believing they can bring their best to the course.

Some instructors will create weekly screencasts instead of sending out a weekly email. This has the particular advantage of being visual. The difficult part is keeping them brief (under five minutes). If you take this approach, I recommend that you still send out the email; include the link to your screencast and a short message.

Chapter 5

What Do I Do with Failing Students? Critical Weeks for Engaging the Bottom 20% of Learners

An after-school remedial reading program in Ohio decided to only accept first graders. Why? All the research pointed to first grade as the most critical year for developing literacy. If students fell behind in the first grade, they would likely never catch up.

Similarly, weeks three and four of an online course have strategic significance. At this point, you now have a few weeks of data on your students- enough to see patterns in their participation. Somewhere between 10-20% of students in an online course will show patterns of absence or disengagement. If this pattern is not addressed early in the course, these students are likely to remain detached the entire semester, and become *Emergency Students*. You know them all too well because they end up creating more work for you, especially at the end of the semester. *But there is a magic bullet for this.*

What these students need is a bit of immediacy, an email from you letting them know that you are aware of their lack of participation in the course.

When students become aware that you can see their site activity and their lack of engagement, they usually will rally and take a new approach to the course. As well, contacting a student sometimes surfaces other issues- a technical problem, some larger family problem, or an illness—giving you an opportunity to connect with the student and to connect them with other school resources: Dean's office, Guidance Counselor, etc.

To gather this information means some technical work on your part. Almost all learning management systems (Moodle, Blackboard, etc.) have reporting features. Browse around on your course administration section and get familiar with these. Lean on your institution's technical support team if you need some help.

What you're looking for:

- Students who have not logged into the course site as often
- Patterns of not submitting course assignments
- Patterns of not accessing critical course media
- Low grades on submitted work

What's Next?

After you've identified the 10-20% of students who are struggling or just disengaged, send them an email. Instead of writing up a new email for each student, you can use this template when you contact them.

Student Name,

I've been reviewing your participation in *name of course,* and noticed that you have not *(items found in the report).* It is an expectation in *name of course* that students will view each week's lectures during the week they are assigned and… *(or other course requirements).* Please let me know if you are encountering some technical issue or if I can be of any assistance with this. (You can access a copy of this text at the *Excellent Online Teaching* site.

An Inoculation Against Absenteeism?

The best long-term solution for online absenteeism is your consistent presence in the course. In general, the courses with the lowest student absenteeism are the courses where teachers are the most present.

Chapter 6

What's the Most Important Habit? Facilitating Discussions

After reading over one thousand student evaluations of online courses, it became evident that the instructors who received the most positive comments were all doing the same thing. It was a single habit, and evidently, the most effective practice for connecting with their students. What were they doing? They were regularly facilitating threaded discussions.

I decided to take a closer look to see exactly what these professors were doing. I was surprised by what I found. I had expected to see professors responding to almost every student post, but that wasn't the case. Often they were responding to just one or two posts per group-- *but that was enough to be remarkable.*

Facilitating threaded discussions isn't just a habit; it's an art. And like with any other skill, it will take you time to develop. What follows are four helpful tips that will get you started.

1. Set a Goal

I frequently get asked the question, "How many times should I post to a discussion?" I recommend that you start with a couple of posts to each group. More importantly, though, set your own goal, and quit when you hit it. It may be to reply to every student that week, or it may simply be to post a follow-up question to the group. Much will depend on your learning objective for the discussion. So, decide on a number and feel good about hitting your own target.

Although you should be aware that it's possible to overwhelm your students with too much involvement (you don't want to shut down student-to-student communication), don't worry too much about it. It's an art, and over time you'll develop a keen sense for the amount of participation your students need from you in a given discussion.

2. Be Brief

The most effective response is often a one or two-sentence question, or a simple request for more information. Done well, this can be an effective tool for inciting deeper learning. Here are a few strategies you could take:

- If a student posted an underdeveloped post, then ask, "Tell us more about why you believe... And what from your reading this week might support this view?"

- Instead of writing a long response to challenge the student's post, ask them to do the heavy lifting. For instance, you might write, "Imagine you are (a position opposite of their own), what would be your two or three arguments against your current position?"

- Suppose a student is missing an important integrative element. You might respond, "How would one's view of ____________ (the topic you think they need to consider) affect the thoughts expressed in your post?"

Such brief replies have an added benefit: they free up time for you to respond to more students. To support this, it's important to communicate to your learners that you expect them to respond to your questions.

3. Timing is Everything

You can imagine what happens when an instructor only replies to discussions at the very end of the week. Only a few students view the instructor comments, and even fewer respond. This is because learners have moved on to the next week's work. It's like last Sunday's newspaper; the readership has disappeared. So, timing is everything. Engage your students' discussion posts during the day when their initial posts are due or on the day after. For example, if the first post is due on Wednesday, reply to them on Wednesday and/or Thursday.

A note on instructor summary posts: These are posts the instructor writes after the discussion is over; in them, the professor sums up the discussion and makes some final comments. Summary posts do have their benefits; however, you run up against the same issue of timing. So, if you write a summary post, send it out as an email to the entire class. This way, you're guaranteed that your students will see it. Better yet, have your students create their own summary posts.

A note on requiring posts: I have strong opinions on this subject. If you want participation, then your discussions will need teeth; that means some form of evaluation. That may mean peer evaluations, grading each discussion, or an overall semester or quarter grade for discussion participation and post quality. Because grading discussions can be more subjective, rubrics become helpful for both the instructor and students. See the resource page for this chapter for links to example rubrics and other resources.

4. Be Clear on Your Purpose.

Before you begin responding to your students, spend some time getting oriented. Read over the discussion prompt and ask yourself, "What's the goal of this discussion?" or "What's the learning outcome I'm trying to facilitate?" This gives you a framework and a focus as you respond.

If you're looking for an area of focus for your own professional development, facilitating threaded discussions would be a great place to spend your time and energy. It's usually the most dynamic and interactive part of an online course, and you can get incredibly creative with these by using role-play, case studies, and student-to-student interviews. In chapters eight and twelve, we'll revisit this topic and add some more tools to your online discussion toolbox.

Chapter 7

How Do I Connect with My Students? 2 Ways to Better Know Your Students

Here's a typical experience from my many years as a commuter student in college and graduate school.

It's after class, and I'm in a conversation with a friend when they mention so-and-so. I nod my head suggesting that I know who they are talking about. But I'm really nodding because I *kind of know* this person…well...I know about him because he is friends with a lot of people I know--but I don't really *know* him. I guess I couldn't even say we were acquaintances. That's the curse of the commuter student: you know about a lot of people, but you don't really get to know them. I've found this dynamic at work in my online classes, too. It's easy to go an entire semester without really getting to know my students.

Here's a quick set of diagnostic questions: If I ran into one of my online students in the student center or walking along the sidewalk, would I be able to greet them by name? Would I know at least one fact about them that would help me begin a conversation? And If I did, how would that impact that student?

Here are a couple of ways to get to know and keep track of what's going on in the lives of your students.

1. Use a Hardcopy of Student Groups or Class Lists

Early in the semester, print off a sheet of each of your groups. If your course is not organized by groups, then print off your entire class list. Before you do, be sure to sort your students alphabetically. Most learning management systems will include your students' photos and email addresses.

Staple these together and put them in a folder or keep them in a three-ring binder.

Then, as you're grading and corresponding with students, use these sheets as a kind of log. You might note something about a student's learning habits, an accommodation, or an extension they may need on an assignment. Sometimes a student will share a recent life circumstance; this is the place to record those kinds of things. Some learning management systems have a feature where you can record this same information within a private notes section. Personally, I find a hardcopy more accessible.

This list becomes useful when you are responding to emails or contacting a student. It familiarizes you with their faces in case you bump into them on campus, or when you meet up for coffee.

2. Checking-in Emails

Another quick way to connect with your students in a more personal way is to send them an email. It can be as short as two sentences: *"Brian, I just wanted to check in with you and see how your semester is going. I'd also like to know if there is anything I can do to make your learning experience a better one. Glad to have you in class this semester."* You can reuse your message by copying and pasting it into several emails (just remember to change the name).

Periodically, click into your participants list and look for students who haven't logged in for several days (you can do this by sorting your students by last login). Send them an email. It may appear that they are not engaging the course, but it could mean that something else is going on. Oftentimes, they are dealing with an illness or a family situation, and are feeling stressed about the course. This is a perfect time to connect with and support them. It takes just a few minutes, but often this makes a big difference in the life of the student.

I once was providing technical support for an instructor who noticed that one of her students had stopped logging into the course. She was new to online teaching, so she contacted me and asked what she should do. I recommended one of these checking-in emails. So, the instructor emailed the student and found out that he was facing a very difficult family situation. The instructor worked with our office and the school's student support systems to help him successfully complete the course. In the process, she discovered that the student lived over 1000 miles away, but was hoping to eventually move to our campus and finish his degree. Because of all that was going on in his life, that dream was now up in the air. However, the care he experienced from his online instructor made it possible for him to succeed. He finished the semester, and his experience in that course became a key factor in his decision to complete his degree. And it all began with a two-minute email.

Chapter 8

How Do I Grade Discussions? Save Time and Increase Learning

I'll be honest, I usually dread grading threaded discussions, mainly because it's so time-consuming. It can feel more like a task to check off than an essential part of the teaching and learning process. But what if there was a way to cut down on your grading time? And, what if you could still provide your students with helpful feedback?

That's where the *Grade-O-Matic* comes in. For just 15 easy payments of $19.99…

Okay…there is no *Grade-O-Matic* – sorry. However, I want to give you a method that I've been using for the last few years, one that's saved me a lot of headache. First, what I used to do: I used to enter grading comments for every student on every discussion. Let's just say that a typical semester had eight discussions, and that a typical class was made up of twenty-five students. That's two hundred grading comments. I'm all for feedback, but what I noticed is that this practice was taking important time away from me, time that I could use to participate in threaded in the threaded discussions. I've noticed the same thing with my colleagues: they spend more time grading discussions than they do facilitating them. So, I want to start by giving you permission to stop. Next, I want to give you a system for grading.

Grading Threaded Discussions – The 20/80 Approach

The 20% - When grading your discussions, only enter comments for the 20% of students who have the lowest scores, the ones who really need some direction for improving their posts. Let them know what they were missing and specifically what they can do better next time. If you are using a rubric (a great way to be clear on expectations), end your comments with, "See the discussion grading rubric for discussion expectations" or direct their attention to a specific part of the rubric that they should review. Finally, invite them to contact you if they need additional help.

The 80% - Create a generic message for the 80% who are doing satisfactory work in their discussions. Copy and paste this message into their grading comments.

Example Message:

Refer to the discussion grading rubric (or syllabus) for the criteria for threaded discussions. I'd also like to invite you to contact me if you have some specific questions about how you can improve your threaded discussion posts, or if you have questions about the topic we have been discussing.

The 100% - Finally, send out an email to the entire class explaining that you have graded all of their threaded discussions. Let them know that you have left specific comments for those who most need improvement, and invite them to email you if they desire more specific feedback on their posts.

This method is based on two things:

- You're investing your energy where you'll be most effective. Instead of spending a ton of time reviewing every discussion and entering grading comments for them, this method frees you to spend more time interacting with your students throughout the week in their discussions. Your in-discussion comments are a more formative method of feedback. Additionally, your in-discussion comments are viewable to the entire discussion group, while grading comments are private and cannot benefit other learners in the course.

- Timely feedback is more important than the quantity or quality of your comments. The more quickly you can get those grades and comments back to your students, the more likely they are to benefit from your comments. This method makes it possible to get the comments to the students who need it, and to get it to them quickly.

In the next chapter, we'll review several practices that should help you make your feedback timelier, not just in discussions, but in your larger assignments as well.

Chapter 9

How Do I Give Feedback? Timeliness & 5 Tips to Make it Happen

If you're a perfectionist, you might not like this next statement: In an online class, prompt feedback is more important than quality feedback.

That's not to say that you can't have both; but, if you are like me, you're familiar with how it goes: sometimes you just don't have that two-hour block of time to focus on grading. During such times, our tendency (at least my tendency) is to put it off until tomorrow. Most of the time, this isn't procrastination; it's just that we want to deliver substantial feedback to our students, and we hope tomorrow will offer us that uninterrupted chunk of time we need. Oftentimes, tomorrow turns into three days, and three days turn into a week, and...

By the time we engage our students, they're on to the next week's reading, a new discussion, or some other assignment. Because of this, we need to strike while the iron is hot, while what they have learned is still fresh in their minds.

Five Practices for Delivering Prompt Feedback

1. Draw a line in the sand.

By setting a date, we prevent ourselves from pushing out our response time in the hope of that uninterrupted space for grading. Decide on how far out is too far out for replying to discussions, or for sending feedback on a particular assignment. Of course, these timeframes will differ based on the nature of the assignment.

2. Set a Regular Time

Residential courses have a set time for class, so why not schedule regular, weekly times for grading, commenting, and responding to emails?

3. Use Tools for Rapid Return

Some technologies are complex and slow down the feedback process; others speed it up. It really all depends on the nature of the assignment and on what works best for you. Are paper and pen your preferred method of grading? If so, let your technical support team know. Your school may have a scanning option for you to use. If that's not available, then consider investing in a good document scanner (see this chapter's resource page for some recommendations) I've found that my students appreciate receiving my handwritten comments when they come to them as a pdf scan of their paper.

One of my favorite tools is *Turnitin*, a plagiarism detection service, and it has a built-in grading program. The grading program is similar to the comment features in *Microsoft Word* - but it's faster. You can even save frequently used comments in a bank, then drag and drop them straight onto the paper scanner. Your institution may have a subscription to this service; if not, you can sign up for an individual account.

4. Use Rubrics to Simplify Grading

Rubrics take some of the subjectivity out of grading. If a student gets 7 out of 10 on a discussion grade, you don't have to write a paragraph describing why they earned that grade; you can simply refer them to the rubric descriptors. This is where that comment box does come in handy when grading discussions.

Chapter 10

What Sort of Feedback Do My Students Need? 3 Strategies for Formative Feedback

You'll remember this feeling from your years as a student: after spending a week or two slaving over an important paper, you finally turn it in to your professor. Then the day comes to get your paper back. You did okay; maybe not as well as you had hoped, but you knew your professor had high expectations. So, you begin flipping through your paper to read the comments. Page one—nothing. Page two—a brief comment about your punctuation. Pages three, four, five, six—nothing. Page seven—something indiscernible, but it appears to be about one of your citations. Your next paper is due in a week, and you're left with little idea for ways to improve.

Student course evaluations are full of versions of these two statements:

1. "I would have liked our professor to interact with us more."
2. "I had hoped for more feedback on my assignments."

They want interaction, a sense of connection. They also want to know what you think about their work and how they can improve. Even the best courses with the best instructors still see these requests for more interaction and feedback. I've driven myself crazy trying to provide what I would hope to be enough. I've shared one part of the solution in chapter nine: timely feedback is more important that high-detail feedback. The second solution is to shift gears from giving summative feedback giving more formative feedback.

Summative vs. Formative Feedback

Summative feedback focuses on assessment; it comes *after* the student completes and submits the assignment. It's performance-driven, and usually involves giving your students a grade. The benefit of summative feedback is that it gives you and the learner a way to measure their learning. For example, let's say you have assigned an essay. You receive all the papers on the due date, mark a grade, leave comments, then students read the returned papers. The drawback to the summative approach is that it tends to be one-way communication (with the exception of the very upset student who sends you the "why did I get this grade" email).

Formative feedback is more of a dialogue; at least it creates space for conversation. Formative feedback focuses on the *process* of learning, so students receive feedback *while* they are crafting their assignment. Formative feedback is less assessment-focused and more learner-focused, engaging the student where they most need to build understanding and skill. Instead of simply assigning an essay, you might implement a peer review where students share their essays and work through a common evaluation rubric. You could participate in these as well, evaluating and commenting on their drafts. Then your students could submit their revised papers for a final grade. You should get better papers—and we all know that better papers are easier to grade.

As an educator, you can see the benefits of taking a more formative approach with your students. In the rest of this chapter, I'll present three ways you can use formative feedback in your online course.

1. Invitation to Dialogue

Students find the online environment new and disorienting. When they had a question in their face-to-face class, they simply raised their hand or talked with you about their paper after class. In person, this just kind of happened. In the online environment, you'll have to create spaces for formative feedback and conversation. Unfortunately, most instructors don't pursue their students in this manner. They expect that their students will contact them with questions. Remember, your students are disoriented and they need some cues. In addition, they are probably not used to taking such an active role in the course. So, you'll have to invite them to interact (*and you'll have to make that same invitation several times*). For K-12 students, I'd even recommend creating a communication/dialogue grade for each quarter.

A few ways to do this:

- During those busy weeks when you know your students are working on a major project, send out an email to let them know you're available to help them. End your email with, "*Simply reply to this email if you have any questions.*"

- If you have certain office hours and are open to taking phone calls, let them know when you are available that week. Be sure to post your phone number on the course site or in the email.

- If you receive a particularly good question from a student, post it as a course-wide email or put it in an Assignment FAQ page.

Whatever strategies you choose, the key is to invite them multiple times, especially at the beginning of the semester. This can help create a culture of dialogue in your course and has significant payoff during the second half of the semester.

2. Set up an All Class Paper Discussion Forum

In this discussion forum, students post their questions about the assignment. They might need help finding credible sources, or it could be that they are having a difficult time understanding a particular concept. If your course site uses groups, I'd recommend turning group mode off for this discussion; that way everyone in the course can see and reply to all the questions that students post. Most discussion boards allow you to subscribe; that way, you get an email alert whenever a student posts a new question. Now, this can get overwhelming and is unnecessary for most discussions, but this forum won't get hundreds of posts like regular discussions. Instead, it becomes a dynamic FAQ page where students can read the responses to the questions they would have asked. Sometimes you'll find that you don't need to reply because another student has already beaten you to it. Because of this, assignment discussion forums can generate meaningful student-to-student interaction.

This can be a real timesaving strategy. If a student emails you with a question that you've already answered in the Paper Discussion forum, you don't have to retype the whole thing; simply send them to the forum. It also gives you a ton of valuable feedback. When you go to redesign or just improve your assignments, the assignment discussion forum will provide you with more helpful and targeted feedback than your course evaluations.

A Quick Tip: To prevent any confusion, it's important to set up separate discussion forums for different assignments.

3. Ask Students for the Feedback They Need

I'll admit that this is more of a summative strategy, but it's one of my favorites because it's a formative take on summative feedback.

Ask your students to type a few questions on the final page of their paper. Their questions may range from confusion about citations, questions on composition, to wondering if their ideas about the subject matter are on track. This gives you a place to start when grading, and it helps you to tailor your feedback to your student's needs. This moves your students into a reflective mode, where they are considering their own learning needs. Additionally, it casts your feedback as a conversation.

Another Quick Tip: To really get this to work, you'll probably need to require a certain number of questions from your students, and for those questions to be part of their paper grade.

If you're relatively new to teaching online, then I'd recommend just picking one of three practices and trying it out this semester. And if you had to pick just one, pick #1, *Invitation to Dialogue.*

Chapter 11

Can My Class Really Become A Community? Tips for Cultivating Connection

I was teaching a hybrid course design seminar and asked some of the participants, "What are the challenges of managing the online classroom?" One of the professors responded with, "There's no personal connection." I didn't respond to his comment, because it wasn't the object of our conversation, but I wanted to say, "You probably didn't cultivate connection." And I also wanted to ask, "How connected is the commuter student sitting in the back of our face-to-face classroom? They come in quietly, listen, maybe take notes, then leave." Whether it's face-to-face or online, nurturing a community of learning is a primary task of the instructor.

I'll admit that I was skeptical when I first heard about all this online community-building stuff. I always have more than enough content to work through and already have a hard enough time accomplishing that. But this term, "learning community," kept coming up, and the research in online learning continued to emphasize this topic of community. Here's the skinny: What studies found was that students who experienced a higher sense of community in their online course were more likely to achieve the learning outcomes. So, I gave it a try -- and it worked.

Here are a few changes I noticed when I began to take this task of community building seriously.

- Students began to contact me more often with questions about what they were learning.

- In discussions, students were communicating more often and with a greater sense of openness to one another.

- I began to receive more emails from students about important personal issues. In the face-to-face context, these were the kinds of things that I might talk with a student about after class or after the school day had ended.

One thing you need to know if you embark on this journey is that community building is an art. So, be patient with yourself as you develop the skills of building community online. The second thing to keep in mind is that you are creating a specific kind of community, a learning community. Your goal isn't to meet students' deep needs for friendship and socializing (though, some of that may happen). The goal is to provide a safe and interactive environment where learning can bud and grow.

Community Building 101

Here are four of the more accessible ways that you can begin building community in your online course:

1. Keep Doing What You're Doing

Instructor modeling is the most important factor in creating a learning community. Your weekly emails, your encouraging remarks to students, your feedback on assignments, and your participation in the course discussions all go a long way in modeling the kind of interaction that's so crucial to for a true learning community.

2. Affirmation Goes a Long Way

Encouragement creates an atmosphere, and it builds a positive sense of community in your course. Catch your students doing something right. Try emailing particular students who have been active in the current week's discussion, and let them know that you appreciate how they are interacting with their group. Oftentimes, students will share personal stories in their posts. Encourage students who demonstrate this kind of openness because it invites other students to be their true selves online.

3. Share Your Life

Share stories from your own experience. Reflect on your own journey in your discipline, what your freshman year was like, where you made mistakes, and what you wish you would have known back then. Share from your weaknesses. A wise professor who I've had the privilege to work with explained that. He said, "People admire you for your strength, but they connect with your vulnerability." As well, share from the present. Take time in your weekly emails or screencasts to share a bit about yourself, what you are reading, or what you're learning (of course, be brief and appropriate with what you share). This kind of immediacy sets the tone for the kind of communication you want to characterize your course.

4. Redesign an Assignment

Sometimes an assignment intended to leverage the power of a learning community just doesn't work, and we need to redesign it. Here's a quick story to illustrate powerful this can be:

I received an email from an online instructor who was ready to scrap his discussions. Why? Students were writing mini-essays for their posts, and they weren't engaging one another.

So, we arranged a meeting and spent about forty-five minutes dreaming up how we could change the assignment.

Here's what the revised assignment looked like:

A. He assigned four book reviews for his history course.

B. Students were given a pool of approved books for each assignment and were given the freedom to choose what book they would read and review.

C. They posted their book reviews online.

D. Two members from their group read their review and then interviewed them about their book.

E. The book reviewer then responded to the interview questions.

It was a success. Here's why:

- The element of choice allowed students to choose a book that caught their interest. This gave students a sense of ownership and motivation.

- It created diversity. In all of the groups, students were reading and reviewing a variety of books, then learning from their classmates' reviews.

- People like to be interviewed. Framing the discussion replies as interview questions was more interesting, and it produced better questions. Because the reviewer was now an "expert" on her book, she was motivated to answer the interview questions.

- Because this professor grouped the assignments by time period, students were reading the reviews with the same historical context in mind. You could do this same thing with a topic, genre, skill, etc.

As learners were now interacting with one another, a vibrant discussion began. Additionally, their discussions were meaningful because they were interacting with the content of the book reviews. At the end of the semester, student evaluations were some of the best I've seen for an online course. A lot of that was due to this assignment and how it effectively leveraged the learning community.

Redesigning an assignment to better incorporate the learning community is a daunting task. If you ask a colleague to assist you in the process, they can bring new ideas to the process and even speed things up. Two-heads are better than one, especially when we're stuck.

Chapter 12

More Ideas for Facilitating Threaded Discussions

You've spent over an hour responding to student discussions. Your eyes are bleary from staring at the computer screen. What's worse is that you still have twenty posts left to grade, and you've run out of things to say. Dr. Scott Wenig, a friend and colleague, has taught a set of online history courses for over ten years. Because students frequently comment that they enjoy his discussion comments, I decided to find out what he was doing. So, I poked into a few of his course discussions. Here's what I found: His tone was always positive, he challenged his students, and he always addressed them by name. More importantly, he was consistent. He did this at the end of the semester, just as he had in the earlier discussions. In short, he had created some powerful habits that sustained the quality of his teaching presence.

We sat down together, and I got a chance to hear exactly what he was doing. Here are the three things he recommended:

1. Create a Rhythm

Break things up and reward yourself for your work. Work for ninety minutes responding to students, then go for a walk, exercise, or grab a hot cup of tea. Tony Swartz, author, and CEO of *The Energy Project,* says that the key to personal productivity is working in 90-minute, focused sessions that are followed by short periods of rest. The method is based on our bodies' physiology, something called the ultradian rhythm.

A few months ago, I was filling out a health survey and was asked how I would rate my workday. It was a painful choice to mark "sedentary." But it's increasingly the reality for those of us who teach online and spend a significant amount of time working at a computer. Not only does taking breaks make us more productive, but it's also just good for our health.

2. Break Things up by Groups

When responding to over twenty discussion posts, things can get blurry. Grading one discussion group at a time helps keep student posts in context, and helps you to perceive group dynamics as you read and respond. This method of batch processing gives you a way to measure your progress and tends to speed-up our grading.

3. Grade When You're Most Alert

For Scott, that's a particular time in the morning; for you, it may be at 10 pm. Whatever time you choose, pick that time when you're able to bring your best.

Seven Go-To Reply Prompts

For those days when you are still drawing a blank on how to respond to your students, here are a few prompts to put in your toolbox. These are one-sentence replies that can lead your students into higher-order thinking.

1. How would you define ____________?

Here you are pulling a significant word from your student's post, one that may be loaded with assumptions that you want them to question. This also demonstrates that you're paying attention to what they are writing.

2. What reasons do you have for saying "______________"?

3. What do you think might be the implications of your statement, "_______________"?

4. What arguments would you expect others to bring against your viewpoint?

5. Tell us more about _________________?

6. How have your views on this changed over the last five years? Why did they change or not change?

7. Share something from your course reading that supports your viewpoint in this post. Explain the connections.

Many times, students share opinions and personal anecdotes, but miss connecting the course content with their work. This is a straightforward way to ask your students to integrate course material into their discussions.

Chapter 13

Four Essential Qualities of an Awesome Online Teacher

How would you like others to describe you to someone else? How would you like one of your students to describe you? Below are four essential qualities of an awesome online teacher.

1. Compassionate

Online, compassion is often shown through noticing. It starts with knowing who your students are (see chapter seven). When we know who they are, and we are patient with them, compassion then has good soil from which it can grow. I'll admit that by year four of teaching online, I had created a very negative portrait of my students. The root of my negativity was that I had not given my classes the appropriate time they needed in my calendar. I expected my students to fit perfectly into my busy week and to not interrupt me with the same questions over and over again. Things shifted when I became more realistic with the amount of time that my courses would require. It also meant adjusting my expectations. Students were going to ask the same questions- the ones I already answered in an email or in the syllabus. Adapting our expectations (without lowering them) makes room for the patience it takes to become an excellent online teacher.

Compassion begins in the soil of patience and connection, but it grows beyond that as well. I think it assumes the best. What I mean is that when a student emails to ask that annoying question, we assume that they are writing because they care about their learning. Whether we're right in our assumption or not, we'll have responded with kindness, and be the better for it. There is something disarming about compassion. The negligent student is often more responsive to kindness than to rigidness. Moreover, compassion doesn't preclude being firm. We can follow our kindness with clear and firm requirements.

2. Good Listener

The excellent online teacher is remembered because of her willingness to listen. But how do you listen online? Have you ever made a hasty reply to an email, only to find out later that you had read the email too fast and missed something important? It's easy to do. Online, listening often means setting aside email until you have time to give it your full attention. As mentioned in the previous chapter, it means working when you are alert, and you can offer your best to your students.

3. Available

When a teacher is available to their students, those students will be more likely to engage in the learning process. There are two aspects to this. First, being available means being reliable; your students don't need 24/7 customer service, but they do need consistency in our replies and feedback. Second, it means being accessible in your demeanor. Your communication shows that you are approachable, responsive, and interested in your students and what they are learning.

If I'm going to be at my computer for an extended period of time, I make it a point to log into my *Gmail* and check my status to display that I'm online. If you keep virtual office hours, this is a great way to make yourself accessible to your students.

4. Flexible

This is a difficult one because there are always those students who, if you give them an inch, will take a mile. We fear that if we push out an assignment date, our students will stop taking due dates seriously. A massive wave of late assignments will then crash onto our desk during the final weeks of the course. The good news is that your course design and syllabus provide you with a lot of structure to lean on, freeing you to be flexible. Sometimes you may need your students to be flexible with you. That was the case for me this semester.

I had crafted a new phased-assignment, but I hadn't taken into account just how demanding it was going to be, nor the very short timeframe my students had left after returning from Spring Break. It included four phases, each with distinct due dates, and culminated in team presentations. A few students sent me emails with a tone that could be described as contained panic. I also picked up that the stress of my unrealistic timelines was straining the students' collaborative relationships. Here's what we did:

First, I apologized for the tight scheduling. Second, I created a chart with the assignments, original due dates, and a column for them to suggest new due dates. Third, I asked for any suggestions on how we could scale back some of the assignments. Finally, I gave them the option of completely scrapping the assignment and returning to my standard curriculum.

What most surprised me was that not one of the teams even mentioned scrapping the assignments. They gave me realistic due dates and helpful feedback. Unintentionally, I had created ownership because my students had been given a voice in the assignment. That ownership has continued throughout the project.

Years ago, I might have thought that my students should just suck it up and get on with their work. After a while, though, that kind of inflexibility just led to strained relationships and stunted learning. Flexibility takes time, but it has powerful effects.

Chapter 14

Communicating Your Unique Voice

One of the reasons we take a class is that we want exposure to a certain teacher. We want to hear their unique take on the subject, and to pick up their passion for their field of study. Students desire learning that is colored with the personality of their teacher. We don't have to lose that online.

Think about why you enjoy certain books. What draws us in, and what makes them easy to read, is the author's voice. I enjoy reading Anne Lammot because she takes me by surprise, and her voice draws my attention to things in life that I would otherwise miss. Chip and Dan Heath, the authors of *Make it Stick* and *Decisive,* write business non-fiction, a genre that most of us read for information; but the Heath brothers' books read like novels, drawing us along from paragraph to paragraph. They are lighthearted, yet serious; well researched, yet humorous.

The Heath Brothers and Anne Lammot, and many other writers have had a profound impact on my life. Have I ever met them? Never. But I've been exposed to their writing voice. Your students need your voice as well. They want to see your subject matter through the lens of your particular story and point of view.

How to Communicate Your Unique Voice Online

- Note your catchwords and catchphrases. Every teacher has these, and if you teach in the face-to-face environment, ask your students about your quirks and catchphrases. Then use them online in your writing.

- Note what you emphasize. What three concepts do you get the most passionate about when teaching your course? Send out three emails over the course of the semester explaining these and why you believe they are so important. We don't want our classes to be a litany of soapboxes; but students remember them, and they are a part of your unique voice. Share them.

- Let them *hear* you. Create screencasts so that your students really get to hear the texture and tone of your voice. I regularly listened to Michael Hyatt's leadership podcast, *This is Your Life*. He also has a blog, as well as several books. Because I've heard his voice, whenever I read his blog or books, I can hear the texture and tone of his voice as I read. It's an incredible function of our brains to do that for us! As mentioned earlier, instead of weekly emails, consider sending your students weekly screencasts.

Check out the online resources for this chapter for screencasting tips and resources. I'll also provide some links to several resources on this subject of developing your voice.

Chapter 15

Improving Your Game

If you've ever been golfing, you understand that it's one of those sports that make you constantly aware of your need for improvement. No matter how skilled you become, you can always come back and shave a few points off your score. Teaching is the same, and the online environment tends to aggravate this dynamic—especially if you're new to it.

As the semester unfolds, we realize just how much depends upon the design of our course. Week by week, we pick up new methods, tools, and we sharpen our communication. This is all good, but it can be overwhelming. It's also easy to lose track of the course revisions we hope to make and the new teaching strategies we hope to employ. In what follows, I want to give you a few tips and tools to help you remember and implement all of those improvements that cross your mind during the course of the semester.

David Allen, productivity consultant to executives, says that we can only remember about five to seven things at a time, and that they all require an amount of mental energy to manage. So, just our daily to-do list maxes out our capacity. Because of this, Allen recommends that we objectify our ideas and tasks simply by writing them down. I highly recommend his GTD (Getting Things Done) system that breaks work down to projects and next actions. Whatever system you have or adopt, the key is getting these ideas recorded somewhere.

One way to get these off your mind is to keep a log. At the top, or on the first page of your course site, you can create a page and call it "improvements for next semester." Keep this hidden from student view. Some learning management systems allow you an instructor notes page, which is ideal. Use this to keep a running record of things you might do differently next semester and course elements you plan to update or improve.

Many of us teach courses that we didn't design. That's okay; good course designers and instructional design teams are hungry for feedback. Yet, I've noticed that many instructors are reluctant to share their observations or recommend improvements to the course design. Keep track of them and share them. What you may find is that there are elements of the course that you can customize and make your own.

Another approach is use a syllabus-focused system. One professor whom I work with reworks his syllabus as he encounters needed changes. He will revise assignment instructions, record notes about what video lecturettes he wants to add, and what elements he plans to eliminate.

A great tool for managing all of this is an application called Evernote. I use this to keep a daily log of what's going well and what's not going so well in my courses. I've switched over to this for a couple of reasons. First, it saves my work in the cloud, so I can access it anywhere. Second, it's searchable, and the search is incredibly fast. (See the resource page for this chapter for details on this tool.)

Consider deploying a mid-course evaluation. Now, this takes some courage, but the feedback you receive will be more valuable than the typical semester-end course evaluation. A graduate professor turned me on to this idea. He was disappointed that the most useful feedback about his course was coming in *after* the course was over. Because some of his courses are taught every other semester, this meant waiting an entire year to make improvements, many of which he was willing to make midstream had he known his students' concerns. I was managing the learning management system; so, he came to me and asked if I could set up a mid-course survey tool. It inspired me to try out this same strategy. The results stirred me to make some of the most effective improvements in my online course. Tap your technical support team if you need assistance setting up a survey for your course, or try using Survey Monkey or Google Forms free services.

Take an online course. Many online teachers have never had the experience of being an online student. Being in the student role is an enlightening experience; the perspective helps in building the compassion that we touched on in chapter thirteen. There are a lot of inexpensive and even free courses to help you improve your game. You'll find several resources for this in the online resource page for this chapter.

Chapter 16

Finishing Well

As a face-to-face course ends, students and professors buzz around campus. You can feel the tension in the air as exams are administered and students carry their papers with them to class, still warm from the printer, and lay them on top of the growing pile on the professor's desk. As an instructor, you have the opportunity to say some concluding words, and in those hurried moments, create a sense of closure. For both the professor and the students, the end of the online semester can be anticlimactic. It lacks that sense of closure we have grown to expect. As well, in our weariness, it's easy to become erratic and hurried in our communication.

Here are four ideas for successfully bringing your course to a close:

1. Takeaway Discussions

Create an optional discussion forum as a place for students to share their takeaways from the semester. This is an excellent way to integrate reflective learning into the course while bringing a sense of closure. Better yet, make this a required and integral element in your course design. Within your discussion prompt, ask students to share how the course has impacted their personal and professional lives. If you work with high school students, ask them to reflect on the "highlights of the course" or what impacted them the most.

2. Employ Culminating Assignments

Culminating assignments can bring a sense of closure to the course and are excellent ways to assess your student's learning. Some examples of culminating assignments are: portfolios, integrative papers, panel presentations, reflection on a real-life experience, timelines, and mini-projects.

Culminating activities can become cumbersome, so consider scaling them down for your students. The power in these assignments is that they tend to move your students into higher-order thinking skills, and contain both reflective and integrative components. Because these require a significant amount of planning and scaffolding, these should be explained and developed early in the course.

3. Concluding Email

Send out a concluding email with your own takeaways from the course. Perhaps you've read a new book that you would like to recommend for further study. You might offer some concluding thoughts on how your students might integrate the subject matter into their own professional development. If you are working with secondary students, explain how the course might benefit them in their future courses and life experience. Share from your own story, and share what you have enjoyed about your students.

4. Host a Q&A Web Conference

This is more technical and takes more prep work, but it can be incredibly effective. Because the last weeks of the quarter or semester are so hectic, I recommend situating your web conference in the third week from the end of the course. Here is one proven format: First, ask (or require) your students to email you questions that they have had during the course. This gives you material to prepare. Second, invite your students to attend the live conference, letting them know that they will be able to share additional questions via live audio or chat. You may want to prepare a few brief slides to share during the web conference in order to add a visual element.

It's important to remember that students take online courses because they are flexible and largely asynchronous. This means that web-conferences are best kept optional. If they are required as part of the course, offer several opportunities for your students to connect. If your learners are adults, this will mean after work hours. Another solution is to record the webinar and make it available to your students after the live session.

5. Plan For It

Now we are back to the idea of leveraging the power of habits and structures. Look over the last weeks of your course. Does it tail off? Or are there elements already built in that require you to interact with your students? If not, consider how you might boost your communication during these weeks. Keep consistent with your weekly emails, and model the kind of engagement you expect from your students.

6. Celebrate

This one is completely for you. This might be dinner out with a friend or your spouse. It might mean enjoying some frozen yogurt. Submit your grades, turn off that computer, and give your eyes a break from that screen.

Conclusion

I remember my superintendent telling me that everything gets better during a teacher's third year. She was right. Then I started teaching online; it was like starting all over again. But her insight held true; with each year, things got better. We get better at our craft, at seeing our students for who they truly are, at developing learning activities that are meaningful (without running ourselves into the ground). My hope is that this succinct book has been a helpful tool to make that happen in your professional life.

This book was designed to grow. So, check back with our online resources often. Our plan is to start with some basics, then to develop the chapter resources further. I would appreciate your feedback on Amazon (it'll be formative).

Finally, feel free to contact me at aaron@excellentonlineteaching.com

About the Author

In 2002, Aaron Johnson began teaching for a private high school in Ohio. When he moved to Colorado in 2005, he decided that he wanted to keep teaching at the school where he had built so many relationships with the staff, faculty, and students. He transitioned from his face-to-face classroom to an online classroom and has been teaching online for them since.

In 2008, with a baby girl on the way, Aaron began working in the Educational Technology department at Denver Seminary, where he continues to work today. As an Associate Dean of Educational Technology, he spends his days helping faculty to create better online courses and to become better online teachers.

Aaron's area of expertise is in creating community and facilitating relationship within the online classroom. He loves watching teachers become confident online teachers, who actually enjoy teaching their online classes. He graduated from Denver Seminary in May of 2013 with an MA in Spiritual Formation. His master's project research explores how instructors can nurture their students' spiritual development through online discussions.

Aaron currently lives in Castle Rock, Colorado with his wife and two amazing daughters. He's an artist at heart, who loves to fly fish and to be in nature. His family loves to hike, play, and go on adventures together.

Online Teaching with Zoom

Book 2 in the Excellent Online Teaching Series

https://excellentonlineteaching.com

Printed in Great Britain
by Amazon